Cape Cod
eco-tales

For more information about Cape Cod Eco-Tales, visit us online:
www.capecodecotales.com

Salt Marsh Secrets

by Heidi Clemmer
illustrated by Marisa Picariello

Tucked safely behind the **barrier beaches** along the bay, are the **salt marshes** of Cape Cod; bursting with species of all shapes and sizes.

Frankie the fiddler crab carefully makes his way out of the **burrow** he has dug with his own walking legs.

He will soon join hundreds of other fiddler crabs for a day on the marsh, but before doing so, he pauses by its entrance...

...taking a moment to wave the larger of his two claws to attract a female fiddler who happens to pass by.

Darting between blades of cord grass which appear to be waving back to him, Frankie skittles through the tiny pools of water in search of something to eat.

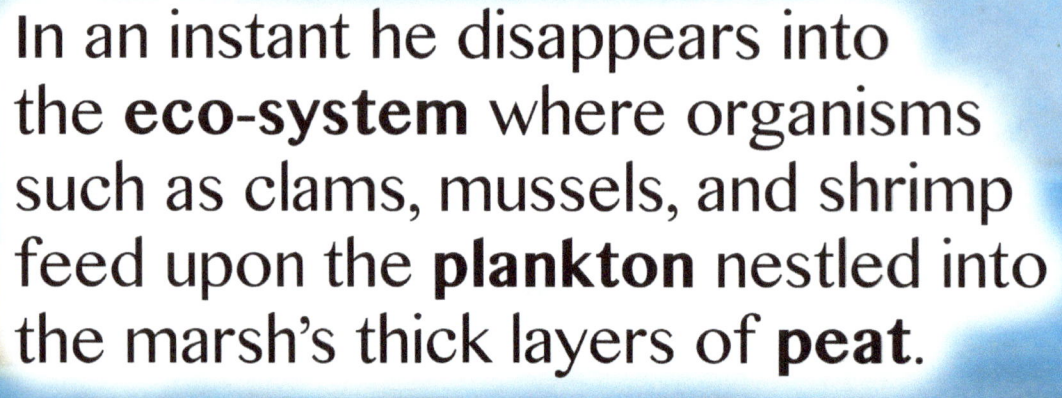

In an instant he disappears into the **eco-system** where organisms such as clams, mussels, and shrimp feed upon the **plankton** nestled into the marsh's thick layers of **peat**.

Tall salt meadow grasses protect Frankie and other salt marsh creatures from wind and high water... and help to hide the nests containing marsh birds, like Eva the snowy egret, from **predators**.

Roland the red-tailed hawk
is perched high upon a scrub pine
at a point where the salt marsh
meets the woods, keeping watch
for any small creatures
scampering along the marsh floor...

...like Frankie
the fiddler crab.

Frankie is now busy fighting other fiddler crabs to defend his burrow, and doesn't notice Roland sweeping suddenly from the sky to snatch him up for dinner.

Meanwhile, Denny the diamond-backed terrapin turtle ventures out to search for a tasty treat the tides may have tossed across the marsh. But Frankie is quicker, and drops into one of the thousands of fiddler crab burrows in the salt marsh to escape both the hawk and the turtle...

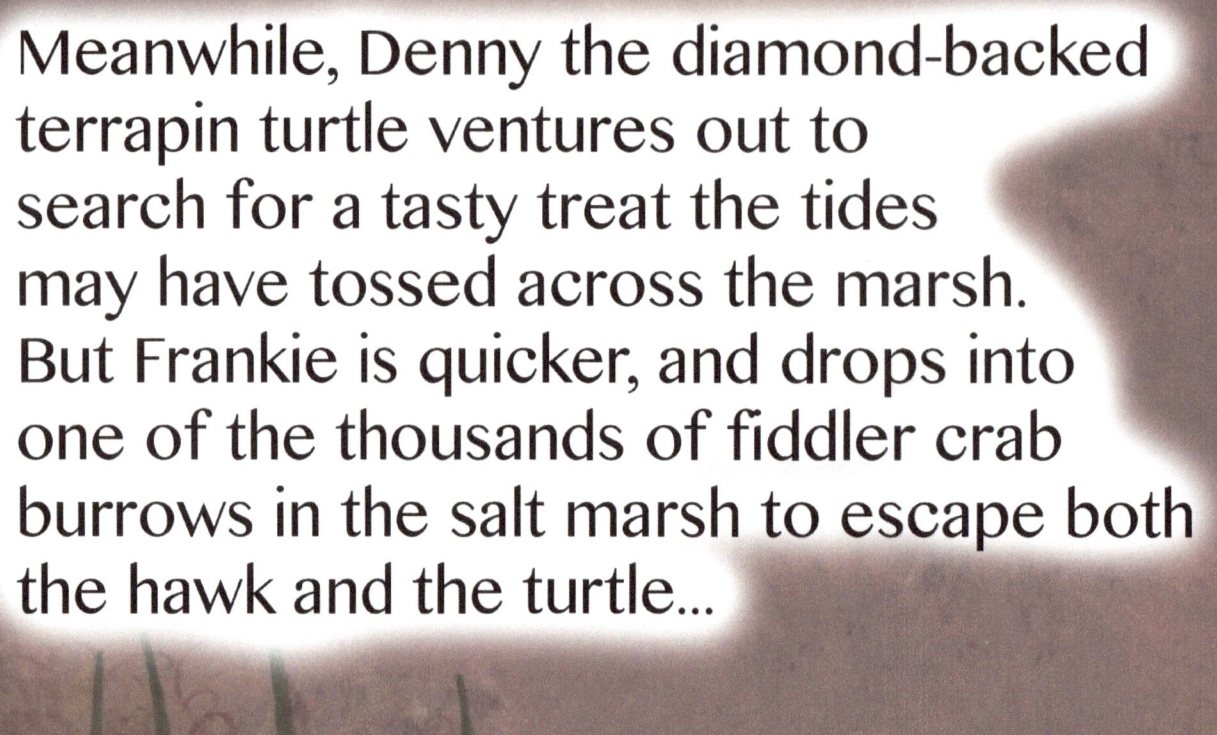

...for today, at least.

Later, as the tide begins to roll back, as it does twice a day in the salt marsh, Frankie begins to plug the hole in his burrow where he will await the next receding tide.

One of the secrets of the salt marsh is that all of the feeding and burrowing that Frankie and the fiddler crabs do in a day, keep the salt marshes clean and help them to grow.

So, if you ever come across this fiddler crab in his salt marsh **habitat** give him a wave and say, "Thank you Frankie!"

Perhaps he'll even wave back!

SALT MARSH

Environmental Characteristics:

Dominant coastal wetlands on Cape Cod;
Occurring behind barrier beaches and divided into two
zones; the high marsh and the low marsh. Driven by the
tides, salt marshes are feeding grounds for numerous
marine species, and function as nurseries for many
commercially important fish and shellfish.

Animal Life:
The Fiddler Crab
The Finfish Crab
The Periwinkle
Shrimp
The Diamond-back Terrapin (turtle)
The Snowy Egret
The Red-tailed Hawk

Plant Life:
Salt-water cordgrass
Salt meadow cordgrass
Salt marsh aster

SALT MARSH GLOSSARY

Barrier beach: a narrow ridge of offshore sand rising just above high tide level and running parallel to the coast

Burrow: a hole or tunnel dug by a small animal

Eco-system: a "community" of interacting organisms and their environment

Habitat: the natural home or environment of an animal, plant, or other organism

Peat: a brown soil-like material formed by partially decaying vegetable matter

Plankton: the small or microscopic organisms that drift or swim weakly in a body of water

Predator: an animal that naturally preys (devours) others

Salt marsh: an area of coastal grassland that is regularly flooded by seawater

THE AUTHOR

Heidi Clemmer was an educator at Wellfleet Elementary School for 21 years. Cape Cod Eco-Tales is the result of her passion for enlightening her students and others to the varied and significant eco-systems that dominate the outer cape landscape.

THE ILLUSTRATOR

Marisa Picariello is a Wellfleet resident who studied art at Wheaton College. She finds inspiration in the natural world and has always enjoyed exploring the Cape Cod landscape.

www.ingramcontent.com/pod-product-compliance
Lightning Source LLC
Chambersburg PA
CBHW050909180526
45159CB00007B/2844